# Borne of the Wind. An Introduction to the Ecology of Michigan's Sand Dunes

 MW00448651

## GREAT LAKES COASTAL DUNES

The high sand dunes are among the most rugged and beautiful natural features of the Lake Michigan and Lake Superior shorelines. Great Lakes dunes comprise the most extensive freshwater dunes in the world, so vast that they are visible to astronauts from outer space. In terms of the plants and animals, the coastal dunes are also rich, supporting more endemic species than any other Great Lakes ecosystem.

Some of the earliest European explorers and settlers in Michigan commented on the rugged character of the dunes. Bela Hubbard, a member of the 1840 survey of the Grand Sable dunes, described them as "a miniature Sahara." Many of Michigan's most visited state parks and national lakeshores showcase these thrilling natural features. Michigan contains more dunes than any other Great Lakes state, however, significant coastal dunes are also found along the shorelines of all Great Lakes states, and on the Ontario shoreline as well.

The high coastal dunes familiar to many Michigan residents are not the only kind of dunes. In fact, other types of dunes may be found along the shoreline as well as in the interior of the state. Here we discuss some of the more common dune types and their significance to a wide variety of plants and animals.

# TABLE OF CONTENTS

Dune Formation & Types ........................................................................1
    Sand Source ...................................................................................2
    Wind ...............................................................................................4
    Water-Level Fluctuation ................................................................5
    Vegetation ......................................................................................6
Major Dune Types ................................................................................8
    Parabolic Dunes .............................................................................9
    Perched Dunes ..............................................................................10
    Linear Dunes ................................................................................12
    Transverse Dunes .........................................................................14
Zonation of Parabolic Dunes ..............................................................16
    Beach ............................................................................................17
    Foredune .......................................................................................19
    Backdune Forests .........................................................................23
    Blowouts .......................................................................................28
    Interdunal Swales & Other Dune Wetlands .................................30
Zonation of Perched Dunes .................................................................32
    Eroding Bluff ................................................................................34
Zonation of Dune & Swale ..................................................................35
    Beach & Foredune .......................................................................36
    Open Interdunal Swales ...............................................................37
    Forested Dune Ridges ..................................................................39
    Forested Swales ...........................................................................40
Zonation of Transverse Dunes within Glacial Embayments ..............41
    Dune Forests ................................................................................42
    Peatlands/Conifer Swamps ..........................................................43
Dune Threats ......................................................................................45
    Exotic Plants & Animals ..............................................................46
    ORVs & Overuse .........................................................................48
    Residential Development ..............................................................50
    Other Development ......................................................................52
Sand Dune Regulations .......................................................................53
Dune Sites in Public Ownership or Private Nature Preserves
    Lower Peninsula ...........................................................................54
    Upper Peninsula ...........................................................................56
Referenced Species .............................................................................58
Exotic Plants Table .............................................................................59
Rare Plants & Animals & Natural Communities ................................60
Recommended Dune Literature ...........................................................62
Selected Scientific Articles .................................................................62

# Dune Formation and Types

Important factors have been identified as critical for the creation of sand dunes, including:

- a source of abundant **sand**

- relatively consistent **wind**

- **water level** fluctuation

- **vegetation** to foster sand accumulation

Dennis Albert

# SAND SOURCE

**The most basic factor in dune formation is the presence of abundant sand.**

Coastal bluffs, a major source of sand for dune formation. Grand Sable Dunes.

Within the Great Lakes basin, continental glaciers covered the landscape for over one million years, providing the major source for sand and other sediments. Meltwater streams flowing from the glaciers' flanks carried tremendous volumes of sand and gravel that settled along extensive sandy plains or in narrow stream channels. Another source of sand was glacial till, a mixture of sand, silt, and clay that was deposited along the margins of the glaciers or dropped as the glacial ice slowly melted. Not only did the glaciers provide sediments rich in sand, they also scoured out the broad, deep basins that the Great Lakes now occupy.

These glacial sediments are concentrated in two sources that provide sand for coastal dunes: rivers and coastal bluffs. Large volumes of sand are carried into the Great Lakes by river systems, which is why many of our dune complexes are located at the mouths of large rivers like the Grand, Kalamazoo, Muskegon, and Pere Marquette. High coastal bluffs, which are continually being eroded by winds and waves, are another important sediment source, as seen at Sleeping Bear Dunes National Lakeshore, along northern Lake Michigan, or at Grand Sable Dunes along the south shore of Lake Superior. These sediments are

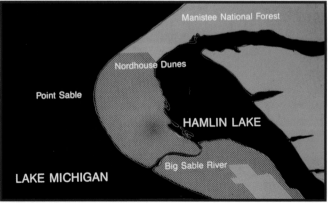

The Big Sable River carried sand that became the Nordhouse Dunes.

moved along the shoreline by offshore currents and eventually get redeposited along the shoreline to form sand beaches or foredunes. When conditions are right, these sands can be carried inland by strong winds to form interior dunes.

While water deposits particles on the beach that range in size from pebbles to very fine sands, wind-formed dunes consist largely of medium and fine sands. Large volumes of medium and fine sands bounce along within a meter of the ground, in a process called **saltation** (see diagram at right). Saltation is responsible for most sand movement on the dunes. Coarse sand is too large to be readily moved by the wind, and its movement consists of short distance "surface creep" resulting from collisions with moving fine and medium sand particles. Finer silt and clay sized particles are carried in suspension far beyond the coastal dunes.

Beach or dune sand consists largely of quartz (87–94%), with lesser amounts of feldspar (10–18%), magnetite (1–3%), and traces of other minerals such as calcite, garnet, and hornblende. Sand particles are rounded and frosted by continuous

collisions with other sand grains. Quartz grains range from very coarse to very fine sands. In contrast, magnetite, which is twice as heavy as quartz, consists primarily of fine and very fine sand grains, although larger particles can be found. As a result of this size and density difference, gently moving water or light winds cause the heavier, smaller magnetite to settle out into dark bands overlain by lighter, larger quartz. Gentle summer winds accelerate the separation of magnetite from quartz, but the sorting is quickly undone

by fall and winter storm winds and waves.

Moving sand is very abrasive, as demonstrated by **ventifacts**, stones that have flattened facets polished smooth by the sand. Similarly, the colorful label of a soda can left on the dune is soon sandblasted to a dull gray. An interesting, rare phenomenon in the dunes is the fusion of sand grains by lightning strikes, forming fragile tubes called **fulgerites.**

Beach sand (enlarged).

David Wisse

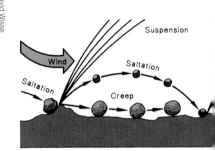

Wind-formed patterns on the dunes. The dark sand is magnetite.

Dennis Albert

3

# WIND

**Strong winds blowing in a relatively consistent direction are a second factor critical for dune formation.**

Wind-blown sand on snow.

David Ewert

For the Great Lakes, prevailing winds are typically from the southwest. As a result, the greatest concentration of large dunes is along the eastern and northern shorelines of Lake Michigan, with the largest dunes along the eastern shoreline. Dune formation is a dynamic and cyclic process. On an annual scale, sand movement and dune formation are greatest from late autumn through early spring, when prevailing winds are strongest and leaves are absent from deciduous shrubs and trees growing on the dunes. Ice accumulation along the shoreline, called an **ice foot**, can reduce the amount of wave erosion of the beach during the winter. During cold years when a large ice foot forms and remains in place throughout much of the winter, there is reduced sand movement and dune formation.

Ice foot along Lake Michigan shoreline.

Donald Petersen

# WATER-LEVEL FLUCTUATION

**Several recent studies indicate that water-level fluctuations influence dune formation, with dune growth accelerated by high water levels.**

Broad beach exposed during low Great Lakes water levels.

 *Dennis Albert*

One such period of intense dune growth was the Nipissing period, occurring about 4000 to 6000 years ago, when Great Lake water levels were considerably higher than today. During this period, the Grand Sable dunes on Lake Superior and the Nordhouse Dunes along northern Lake Michigan were formed. It was once thought that all of the dunes along the eastern shore of Lake Michigan were formed during Nipissing time, but recent studies have shown that some of the Lake Michigan dune complexes, such as those near Muskegon and Grand Haven, have formed as recently as 3000 years ago, during later high-water periods. If we look at dune formation over a period of several thousand years, we see that there were periods of intense dune formation, followed by periods of reduced dune activity.

Dune formation appears to be linked to high water levels in the Great Lakes, when waves and storms erode away greater amounts of sandy sediments than when water levels are low. Relatively high water levels result in destabilization of the dunes, with increased sand movement and burial of forests. Thus, when researchers or sand miners dig deep into a dune, they often encounter several buried forests, one on top of another, separated by

dune sands. These forests established and grew during low water periods of the Great Lakes, and were then buried and killed during a subsequent period of high water, when the dunes destabilized and began to move. Great Lakes water levels vary cyclically, with minor cycles of five to ten years, and more major cycles of thirty to several hundred years.

Northern white-cedars buried by sand.

Wave erosion of foredune during high water levels.

*Dennis Albert*

*Donald Petersen*

# VEGETATION

**A final factor, vegetation, traps and stabilizes sand.**

Donald Petersen

$W$ithout periodic entrapment, sands cannot accumulate vertically to create high coastal dunes. Most plants are not adapted to the constant burial and abrasion that characterizes the dune environment.

Narrow spikes of marram grass *(Ammophila breviligulata)* and broad spikes of sand reed grass *(Calamovilfa longifolia)* colonize open sand.

Susan Crispin

Buried cottonwood *(Populus deltoides)* produces abundant root suckers.

Marram grass (*Ammophila breviligulata*), typically the first species to establish on the bare dune sands, is one of the species most adapted to survival on the dunes. Not only can marram grass tolerate being buried by sand, studies have shown that it requires burial for optimal growth. When the wind encounters marram grass or other dune vegetation, its velocity is reduced, causing sand to accumulate. As dune sands bury the marram grass, it continues to form new growth above the sand, while its roots and rhizomes continue to grow and stabilize the sands.

Donald Petersen

Marram grasses slow the wind, causing sand to accumulate.

Most of the common herbs, shrubs, and trees of the dunes are tolerant of sand burial. Grasses most tolerant of sand burial include marram grass, sand reed, and little bluestem. Shrubs tolerant of burial include red osier dogwood and sand cherry. In southern Michigan, cottonwood shows considerable tolerance to sand burial, as does balsam poplar further to the north. When buried, both cotton-wood and balsam poplar produce root suckers from buried stems, resulting in clumps of small trees on the partially stabilized dune. Roots form just below the sand surface, where moisture is most available.

As vegetation of the dunes is stabilized, herb and shrub diversity increases and forests eventually establish. With the establishment of a forest canopy, dune pioneering plants like marram grass and sand cherry rapidly disappear and are replaced by plants more tolerant of the reduced light conditions in the forest understory.

Dennis Albert

Little bluestem
(*Schizachyrium scoparius*).

# Major Dune Types

**Classification and Distribution of Major Dune Types**

There are four distinctive types of dunes encountered in Michigan: **parabolic**, **perched**, **linear**, and **transverse**. The first three dune types are commonly associated with the present Great Lakes shoreline, while transverse dunes are more often associated with large bays of Glacial Lake Algonquin, from about 11,000 years ago. Linear dunes have also been called **dune-and-swale complexes**. Several classifications of sand dunes identify several more categories than the four listed here. Some of these studies are listed at the end of this publication for further reference.

While the shape of each dune type differs, the general cross-section of all dune types has similarity. The slope on the windward face of a dune is gentle, usually not over 15 degrees. In contrast, the slope of the back or lee side of the dune is much steeper, sometimes approaching the "angle of repose" of dry sand. The lee slope is steep enough that climbing it can be exhausting and typically results in considerable erosion of the slope.

# PARABOLIC DUNES

**Parabolic dunes, defined by their distinctive U-shape, are found only in moist environments, where extensive vegetation cover often stabilizes the dunes.**

R. Torresen

Parabolic dune at Hoffmaster State Park extends into the surrounding forest.

Parabolic dunes occur in several large complexes along the eastern Lake Michigan shoreline. In the Lake Michigan basin, the sand source for the parabolic dunes is broad sand terraces that formed along the lake margins 11,000 to 13,000 years ago. Parabolic dunes probably formed when the forested sand terraces were destabilized during high lake levels, resulting in the formation of U-shaped blowouts. Dune destabilization occurred at intervals several hundred or thousand years apart, when lake levels were high. Strong winds result in migration of the dunes inland, while marram grass and other vegetation cause the sand to accumulate vertically. The end result is the formation of parabolic dunes 250 to 300 feet high.

IMARY WIND DIRECTION

# PERCHED DUNES

**These dunes are perched atop glacial moraines that have bluffs 90 to 360 feet above the present lake level.**

Perched dunes are restricted to the northeastern shore of Lake Michigan and to Lake Superior. They are nourished largely by sand blowing off bluff faces, rather than sand from the beach. Perched dunes themselves are often much smaller features than the bluffs upon which they rest. Foredunes, perched along the upper edge of bluffs, are quite dynamic. Among the most famous perched dunes are Grand Sable on Lake Superior and Sleeping Bear dunes on northeastern Lake Michigan.

Susan Crispin

Grand Sable Dunes, perched dunes along Lake Superior.

Sleeping Bear Dunes, perched dunes along Lake Michigan.

A lag zone of pebbles and cobbles along the bluff-top at Grand Sable Dunes.

The sediments of bluff faces are quite variable. Clay banding can result in moist seepages on the bluffs. Strong winds remove clay, silt, and sand, leaving pebbles and cobbles, called a **lag zone**, either on the face of the bluffs or along the upper edge. Along the lakeshore, the beach is often quite narrow.

Susan Crispin

Gary Reese

# LINEAR DUNES

**Large complexes of linear dunes form the shoreline along numerous Great Lake bays.**

Aerial photograph of a dune and swale complex along northern Lake Michigan.

Ted Cline (Photair, Inc.)

These complexes consist of a series of roughly parallel dunes that form as the water level of the Great Lakes gradually drops. There have been several names applied to these complexes of low dune ridges, including linear dunes, dune and swale complex, wooded dune and swale complex, shore-parallel dune ridges, and beach-ridge complex. We use the terms **linear dune** and **dune and swale complex** in this publication.

Several recent studies have been conducted on Lake Michigan's dune and swale complexes, analyzing the geological form and function, history of lake-level fluctuation, and changes in vegetation over time (succession). The lake-level studies document that most of these broad complexes date from the Nipissing Phase of the Great Lakes, about 4000 to

5000 years ago. Since this time, Great Lakes water levels have been gradually dropping over the long term. The water levels also have an irregular short-term cycle of lows and highs. As the water level drops, a new beach ridge (foredune) forms along the lakeshore, usually separated from the previously formed ridge by a swale. The largest beach ridges in a dune and swale complex record periods of high water level. When the water level rises high enough, beach ridges and swales formed at lower water levels will be destroyed, leaving no record of their short existence.

The large dune and swale complex at Indiana Dunes National Lakeshore on Lake Michigan in Indiana contains upwards of 150 dune ridges, forming a four-mile wide complex that was deposited over roughly 4000 years. Along northern Lake Michigan, at Michigan's Wilderness State Park, a two-mile-wide complex of 108 ridges was created over a similar time period. Unlike the parabolic dune complexes, most dune ridges in the dune and swale complexes are low features, generally less than 15 feet in height. However, within many dune and swale complexes, there are periods when water levels were high enough to destabilize the small dunes, forming small groups of parabolic dunes within the larger parallel dune and swale complex.

Along the shoreline, plant communities are similar to those found on the beach and foredune of parabolic dunes. Within 2 or 3 low dune ridges from the shore, forests have typically established. Usually only the first or second swales along the Great Lake are flooded and dominated by marsh vegetation. Farther inland, most swales are drier and support shrub swamp or treed swamp. This may be due to water level drops, rebound of the land following melting of the mile-thick glacial ice (called isostatic rebound), or a combination of both factors. Along Lake Superior, where this rebound is greatest, swales are quite dry, often supporting jack pine or red pine forests.

Dennis Albert

Red pine and white pine grow on the dunes surrounding a shallow, open swale near Gulliver on northern Lake Michigan.

# TRANSVERSE DUNES

**In the Upper Peninsula of Michigan, large abandoned bays of glacial Great Lake Algonquin from approximately 11,000 years ago contain transverse dunes and an occasional parabolic dune.**

Transverse dunes, linear to scalloped in shape, are believed to have formed in shallow bays along the edge of the glaciers. Sand carried by glacial meltwater streams was abundant along the margins of the bays. Strong winds blew off the glaciers, forming this sand into a series of long, linear dunes, all oriented roughly perpendicular to the wind. The dunes have steep south faces, indicating that the winds were from the north. They are 30 or 60 feet high and are surrounded by shallow peatlands. Many of the Algonquin-aged bays were large, greater than 5 miles across, and contain several dozen dunes. Similar complexes of transverse dunes are also known from portions of northern Europe. Some scientists still debate the origin of these sand ridges, thinking that these dune features are actually sandspits that formed in shallow water. Within the dune complexes there are also localized parabolic dunes.

Transverse Dunes,
Ogontz Bay, Delta Count

For several thousand years after the glaciers retreated, the climatic conditions were dry enough that pine forests grew on the flat sands of the dried-up bays, but about three to four thousand years ago, the climate became cooler and moister, allowing wetland plants to expand across the bays, replacing pines with aquatic sedges and sphagnum mosses and creating peatlands that now have thin peat deposits, 2 to 3 feet thick.

Dennis Albert

Small transverse dunes within a vast peatland in northern Luce County.

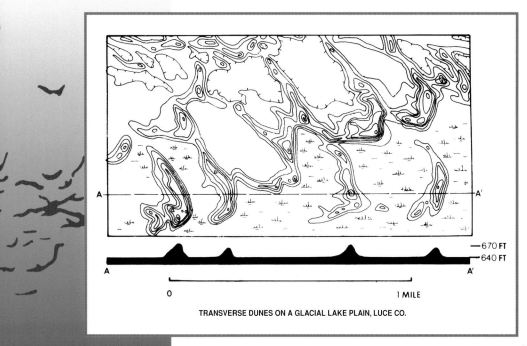

A ———————————————— A'

—670 FT
—640 FT

A                          A'

0                              1 MILE

**TRANSVERSE DUNES ON A GLACIAL LAKE PLAIN, LUCE CO.**

# Zonation of Parabolic Dunes

Photo courtesy of Michigan Natural Features Inventory

Piping plover *(Charadrius melodus)* nesting on a gravel beach.

All dune features have distinct zones determined largely by the physical processes of dune formation: transport of sand along the shore by waves and currents, followed by wind transport of sand to create dune features. The characteristic zones of parabolic dunes are the **beach**, **foredune**, **interdunal wetland** or **trough**, and the **backdune**, each with its distinctive physical character and biota. The sharp physical contrasts between these zones, combined with the dynamic nature of the coastal environment, has led the dunes to be a focus of many studies of vegetation change over time (called **succession**), with the classic ecological studies of Henry Chandler Cowles in the 1890s and J. Olson in the 1950s among the best known. Pioneer herbaceous plant species establish on the bare sand of the beach and foredune, and are gradually replaced by shrubs and trees as soil moisture and nutrient conditions improve on the backdune over time.

Earl Wolf

Marram grass stabilizes a low foredune behind a sand beach. Hoffmaster State Park.

**The beach is the most dynamic zone, where wind, waves, and coastal currents create an ever changing environment.**

Daily and seasonal extremes occur in moisture conditions, substrate, and temperature. Wind velocity is highest at the shoreline, as is the frequency of flooding and the wave energy. Because of these extremes, few plants and animals are adapted to live on the beach. Scattered plants of sea rocket, an annual, are found growing near the water's edge. Any plant that establishes on the beach during the summer is likely to be torn from the shifting sand by powerful winter storm waves, giving annual plants a strong advantage over more long-lived perennials.

Many animals visit the beach to feed. Shorebirds, including gulls, sanderlings, sandpipers, and plovers, feed on drowned insects and dead fish along the beach, especially during spring and fall migrations. Caspian terns, nesting in protected areas along the shoreline, can be seen diving for fish near the shore. Storm waves move debris, aquatic-plant remains, dead insects and fish, and driftwood far up the beach, creating a zone called the storm beach, where flies, tiger beetles, and other insects can often be found in great numbers feeding.

Beach pea (*Lathyrus japonicus*).

Earl Wolf

Sea rocket (*Cakile edentula*).

Earl Wolf

Powell Cottrille

. . . from zone to zone

LAKE | BEACH | FOREDUNE | TROUGH | BACKDUNE

Tiger beetle (*Cicindela scutellaris*).

Occasionally groups of butterflies, like the white admiral, can be seen puddling (another term for drinking) near the shoreline, where evaporation causes minerals to concentrate in the moist sand. Minerals are lost by butterflies during breeding and the moist beach sands provide the butterflies with a nutrient-rich moisture source.

Many species of tiger beetle are found along sand beaches of the coastal dunes. During migration, thousands of ladybird beetles can be found clustered together on debris along the shoreline. These are but a small sample of the insects that use the beach.

Further up on the beach, where storm waves seldom reach during the summer months, plants tolerant of the strong winds and extreme temperatures can establish. These include sea rocket, beach pea, bugseed, and seaside spurge, most of which are annuals adapted to a harsh, changing environment. Digger wasps are also found in this zone.

Where the back beach has abundant pebbles, piping plover, a federally endangered shorebird, creates a nest of pebbles to raise its young. Piping plover was once common along Great Lakes beaches, with 30 nesting pairs once counted along two miles of southern Lake Michigan beach. Market hunting in the late 1800s greatly reduced its populations, but more recently it has again been threatened by increased human development along the shoreline. Raccoons and unleashed dogs both destroy plover eggs and kill juvenile birds.

White admirals (Limenitis arthemis) and tiger swallowtails (Papilio glaucus) puddling on beach.

Powell Cottrille

**The foredune is the zone where pioneering grasses cause sand to begin accumulating.**

Marram grass (*Ammophila breviligulata*).

Earl Wolf

It can consist of a distinctive, low dune feature, or merely a low landward continuation of the beach. The most important pioneering grass on the foredune is marram grass, which is among the plants most adapted to sand burial. Each spring, it sends a thick rhizome toward the surface of the windblown sand before its stiff leaves begin to elongate. Its dense fine roots trap sand particles, helping to stabilize the blowing sand. Sand reed, another dune grass that stabilizes dunes, typically establishes after marram grass has arrived. Among the other plants that tolerate sand burial are several shrubs, including sand cherry, red osier dogwood, bearberry, dune willows, and creeping juniper. Not only must plants of the foredune tolerate periodic burial, they must also survive extreme temperatures, low moisture levels, and low levels of nutrients. Some of the herbaceous plants adapted to these conditions are hairy puccoon, beach pea, wormwood, harebell, and common milkweed.

Plants found growing only on the beach or foredune depend on the extreme conditions for their survival. Two of these are Pitcher's thistle and Lake Huron tansy. Without the open, blowing sands, these Great Lakes endemic plants would be unable to compete with other plants. Probably their greatest threat is the establishment of exotic (non-

Pitcher's thistle (*Cirsium pitcheri*).

Susan Crispin

Harebell (*Campanula rotundifolia*).

Photo courtesy of The Nature Conservancy.

Pitcher's thistle *(Cirsium pitcheri)*.

Antlions (Family *Myrmeleontidae*) are typically hidden beneath the sand.

native) plants that effectively stabilize the dune sand.

During the summer, temperatures at the sand surface regularly reach 120° F (50° C), locally reaching 180° F (80° C). Such extremes cause many of this zone's inhabitants, such as Fowler's toad, the eastern hognose snake, spider wasps (also called digger wasps), sand spiders, burrowing spiders, and wolf spiders to burrow during the day to reach cooler temperatures. When temperatures are extreme, they are active at the surface only during cooler morning, evening, and nighttime hours. The antlion is another insect that spends its life under the sand, waiting for ants and other small insects to slip into its unstable funnel-shaped sand trap.

The harmless eastern hognose snake is one of the most intriguing residents of the dunes. When surprised, the hognose snake rises up defiantly, flattening its head into a cobra-like hood and hissing loudly. If this threatening posture fails, it will writhe, convulse, and finally turn over on its back and feign death, only to flip over again and continue on its way once danger has passed. Toads are a favorite food of the hognose snake. Another large and extremely fast snake occasionally sighted sunning itself on the dunes or in low shrubs, is the blue racer.

A rare insect of the foredune is the Lake Huron locust. This locust exposes its yellow wings and clicks distinctively during flight,

Wolf spider (Family *Lycosidae*).

Sand cherry *(Prunus pumila)*.

Hairy puccoon *(Lithospermum caroliniense)*.

Earl Wolf

Eastern hognose snake *(Heterodon platirhinos)*.

providing means of easily recognizing an otherwise indistinctive insect. The locust feeds on wormwood, a common plant on the dunes. Wormwood is also the host for a rare plant of the dunes, clustered broom rape, which is partially parasitic on wormwood. A rare moth found only on the foredune is dune cutworm, whose larvae feed on dune grasses. Song sparrow, vesper sparrow, field sparrow, and the eastern bluebird forage on insects of the foredune, as do eastern kingbird and chipping sparrows when scattered trees are present. Prairie warbler, a rare breeding bird in Michigan, is known to nest among

Earl Wolf

Eastern hognose snake *(Heterodon platirhinos)*.

Ken Jacobsen

Red anemone *(Anemone multifida)*.

Gary Reese

Lake Huron tansy *(Tanacetum huronense)*.

Photo courtesy of The Nature Conservancy

Szterile rosette of Pitcher's thistle (Cirsium pitcheri).

the shrubs on and in the lee of the foredune, as far north as Sleeping Bear Dunes along Lake Michigan and Rogers City on Lake Huron. Colonies of bank sparrows nest in the low, steep cliff face of the foredune, formed by erosion during winter storms, and both bank swallows and tree swallows are often seen feasting on the abundant flying insects of the dunes.

The foredune is continually being buried by sand or eroded by strong winds. It is colonized by the beach grasses, along with other herbs and shrubs tolerant of sand burial, but its environment is typically too extreme for the successful establishment of most tree species. A few eastern cottonwoods, balsam poplar, and oaks can be found on the most protected portions of the foredune. All of these trees, especially the eastern cottonwood and balsam poplar, tolerate burial by sand, but the strong winds and occasional burial result in short stature and poor growth.

Dune cutworm (Euxoa aurulenta).

Butterfly weed (Asclepias tuberosa).

Spider wasp (Pompilidae).

Bearberry (Arctostaphylos uva-ursi).

Fowler's toad (Bufo fowleri).

Ground juniper (Juniperus horizontalis).

David Cuthrell

Earl Wolf

David Cuthrell

Earl Wolf

Earl Wolf

Photo courtesy of The Nature Conservancy.

# BACKDUNE FORESTS

**Eventually dune grasses, herbs, and shrubs stabilize the sand and larger dunes form behind the foredune, protected from some of the force of the wind.**

Sand deposited during the winter on the lee side of a parabolic dune.

Donald Petersen

During high water periods, dunes continue to grow vertically as they move inland. Behind the protective foredune, winds are lighter and sand accumulation is slower, allowing trees to establish and form forests. On the wind-exposed dunes closest to the shoreline, oaks, bigtooth aspen, and sassafras grow on the dry, nutrient poor sand. Dark, irregular crowns of white pine reach through this open forest of deciduous hardwoods. This remains a harsh environment, where storms regularly uproot trees and large amounts of sand can be deposited, especially during the fall and winter.

Large numbers of monarch butterflies can be seen along the coast during spring and fall migrations. While they are most often seen flying on the open foredune, large clusters of monarchs seek thermal protection on the branches of an oak or cottonwood within the open dune forest, where they can sometimes be seen in the

evening or early morning. Monarch butterflies are not alone in their use of the shoreline as a migration corridor. Other migrants include songbirds, hawks, nighthawks, and owls. Many species of raptor take advantage of the strong coastal winds during their migration. It is not uncommon to see hundreds

Monarch butterflies *(Danaus plexippus)* congregated on a black oak along the edge of the dune forest.

Photo courtesy of Michigan State Parks.

of broad-winged hawks gliding from one warm air thermal to another, saving energy during their long migrations to and from South America.

Open stands of jack pine, sometimes referred to as Great Lakes barrens, occur commonly on the exposed, windward slopes of many dunes. Jack pine and common juniper tolerate the droughty conditions, as do bearberry, creeping juniper, and other shrubs. It is not uncommon to see jack pine growing in the shallow water of the interdunal swales, where it likely established during much drier conditions.

Common trillium *(Trillium grandiflorum)* in the backdune forest.

In contrast to the exposed windward slopes, wind velocity is much reduced on the landward (lee)side of the dunes. Forest trees are protected from severe storm winds and extreme levels of sand deposition. Over hundreds of years, organic material accumulates on the forest floor, increasing the amount of available moisture and nutrients, allowing mesic forest species like sugar maple, American beech, basswood, and red oak to establish. Eastern hemlock forms dense stands in many of the most sheltered, steep coves between dunes, where there is shelter from strong winds and the dense forest

Eastern box turtle *(Terrapene carolina carolina)*.

Hepatica *(Hepatica americana)*.

canopy reduces the summer heat and moisture loss.

But even in this sheltered environment, fine and very fine sand are blown far inland by strong winter winds and deposited as a film on the snow. These fine sand particles, which appear to contain a greater amount of nutrient-rich calcite and feldspar than typical dune sand, may add both nutrients and moisture retaining capacity to the forests behind the high parabolic dunes.

The rich forest floor supports a diversity of spring wild flowers, including common and nodding trilliums, jack-in-the-pulpit, squirrel corn, yellow violet, bunchberry, spring beauty, hepatica, Solomon's seal, and bishop's cap. Still many more wild flower species grow in the backdune forest. In the backdune forests along southern Lake Michigan, southern species such as prairie trillium are found growing at the northern edge of their range. Most southern species are not found north of Muskegon, where a more northern flora appears.

Animal diversity is also high in the backdune forest. Forest songbirds are abundant and

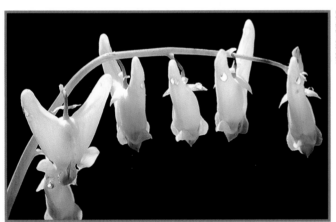

Donald Petersen

Dutchman's-breeches (Dicentra cucullaria).

Earl Wolf

Jeff Dykehouse

Betty Cottrille

Star flower (Trientalis borealis).

Wood thrush (Hylocichla mustelina).

Jack-in-the-pulpit (Arisaema triphyllum).

Solomon's seal
(*Polygonum pubescens*).

Bunchberry (*Cornus canadensis*).

Prairie phlox (*Phlox pilosa*).

Blue-spotted salamander (*Ambystoma laterale*).

include black-throated green warbler, black-and-white warbler, scarlet tanager, wood thrush, ovenbird, eastern wood pewee, rose-breasted grosbeak, veery, American redstart, yellow-billed cuckoo, black-billed cuckoo, and red-eyed vireo. This abundance of songbirds attracts many raptors, such as sharp-shinned, red-shouldered, and Cooper's hawks. The eastern box turtle can also be encountered ambling

Canada anemone
(*Anemone canadensis*).

Trout-lily (*Erythronium americanum*).

Squirrel-corn (*Dicentra canadensis*).

Earl Wolf

Red-backed salamander
(Plethodon cinereus).

Earl Wolf

Bishop's cap (Mitella diphylla).

Earl Wolf

Rue anemone
(Anemonella thallictroides).

on the forest floor, and careful investigation beneath logs and leaves will sometimes uncover the delicate red-backed salamander. The casual visitor seldom encounters more than a few red-backed salamanders during a visit, but this small salamander may be more important to the forest than many realize. Studies have shown that from 200 to 3600 salamanders can live in an acre of forest.

The salamander feeds on a broad range of insects and soil organisms, as well as being eaten by birds, turtles, and small mammals of the forest. The black rat snake is also a localized inhabitant of the mesic backdune forest along southern Lake Michigan.

While the backdune forest is the most protected habitat on the dunes, it can also undergo dynamic change. Its location

along the shoreline results in exposure to infrequent, but extreme storm events. Such coastal storms during the spring of 1998 destroyed backdune forests at Hoffmaster State Park and Wilderness State Park in Michigan. Long-term studies have been initiated to learn more about the successional changes to the forest resulting from these storms.

Earl Wolf

Bloodroot (Sanguinaria canadensis).

Earl Wolf

May-apple (Podophyllum peltatum).

Earl Wolf

Spring-beauty (Claytonia virginica).

# BLOWOUTS

**Blowouts are U-shaped areas of open, migrating sand that occur on dunes otherwise stabilized by forest vegetation.**

R. Torresen

Blowout in parabolic dunes, Hoffmaster State Park.

Learn to recognize the three part leaf of poison ivy *(Toxicodendron radicans)*.

Earl Wolf

Studies show that there have been several periods during the last 4,000 years when environmental conditions resulted in extensive blowouts within the dunes. These periods are closely linked with Great Lake high-water levels.

Dune grasses partially stabilize the blowouts resulting in vertical accumulation of sand within the blowout. As the sand migrates inland, it buries and kills the forest. Sand migration is greatest during the fall and winter months, when vegetation cannot effectively reduce the wind velocity. Most of the plants and animals characteristic of foredunes are also found on blowouts.

Pitcher's thistle can be found growing throughout the bowl of a blowout and is often quite abundant near the upper, landward edge, where there is abundant open sand. Lake Huron locust can be abundant here as well. Poison ivy can be especially troublesome along the lee edge of the blowouts, where it sometimes forms a formidable, waist-high wall between the forest and the blowout.

**Small ponds are common between the foredune and the backdune in many dune complexes.**

Maureen Houghton

Interdunal wetland, Muskegon State Park.

James Harding

Blanchard's cricket frog
(*Acris crepitans blanchardi*).

In these ponds, water levels are controlled by the Great Lakes; during low lake levels the ponds may be completely dry. As a result of these water level fluctuations, both the plant and animal life can change dramatically from year to year. Rushes and bulrushes are often common in the shallow, warm water. Jack pines are also common along the margins of the ponds and occasionally grow within sphagnum mosses in these wetlands. In a few southern Michigan coastal wetlands, there are plant species typically found growing along the coastal plain of the Atlantic Ocean. These species growing outside their normal growing range are called **disjuncts**. Many coastal disjuncts respond dramatically to water-level fluctuations, sometimes disappearing when water is deep, and forming a carpet of plants when the ponds are dry. Houghton's goldenrod is another rare plant that grows only along the moist margins of these Great Lakes coastal wetlands in northern Michigan. When water levels drop, the underground stems (rhizomes) of Houghton's goldenrod expand out into the wetland, where they produce many more flowering stems, only to contract again when the pond expands in later years.

Gary Reese

Aerial photograph of interdunal wetlands at Ludington State Park.

Earl Wolf

Spotted turtle (*Clemmys guttata*).

Gary Reese

Northern leopard frog (*Rana pipiens*).

# OTHER DUNE WETLANDS

The shallow ponds are rich in aquatic insects like dragonflies and damselflies. Spring peepers, Fowler's toads, garter snakes, mink, and muskrats are among the other inhabitants of the ponds. Along southern Lake Michigan, the rare Blanchard's cricket frog occurs locally in some shallow interdunal ponds. Spotted sandpipers build nests around the ponds, where there is greater protection than on the nearby exposed Great Lakes beach.

While most of the ponds are found between the foredune and backdunes, there are often large swamps and ponds located at the inland edge of the dune fields. These shallow ponds, with accumulations of organic material on their bottoms, support an even greater diversity of life than the smaller interdunal ponds. Migratory songbirds take advantage of the wetlands' thermal protection and feed heavily on the abundant midges and other aquatic insects during spring migration, when other sources of forage are scarce.

Maureen Houghton

Swamp forest with marsh-marigold *(Caltha palustris)* along the back side of the dunes, Hoffmaster State Park.

Earl Wolf

Jack pine *(Pinus banksiana)*.

Gary Reese

Virginia meadow-beauty *(Rhexia virginica)*.

David Wisse

Interdunal wetlands bordered by jack pine, Ludington State Park.

# Zonation of Perched Dunes

Perched dunes share the **foredune**, **blowout**, **backdune forest, and interdunal wetland** found on the parabolic dunes, and most of the discussion of these zones under parabolic dunes is relevant to the perched dunes as well. Many of the rare plants and animals of the parabolic dunes, including Pitcher's thistle, Lake Huron tansy, clustered broom rape, and the Lake Huron locust are also common within the perched dunes. Two rare plants that are especially common on the perched dunes are dunewort and clustered broom rape, both disjunct from the western United States.

Gary Reese

Perched dunes, Grand Sable Dunes.

Dennis Albert

Mosses and earth stars (*Geastrum* sp.) growing on open dunes, Grand Sable Dunes.

Dennis Albert

Wormwood *(Artemisia campestris).*

Mosses appear to play a role in dune stabilization on some northern dunes. This is most evident on the Grand Sable dunes along Lake Superior, in the Upper Peninsula of Michigan, where cooler, foggy conditions are more favorable to mosses and lichens. Studies in the Grand Sable dunes showed that *Polytrichum piliferum*, a moss, was able to survive burial by over two inches of sand.

The perched dunes themselves are by no means stable. The largest perched dune at Sleeping Bear Dunes National Lakeshore, the "Bear", was completely forested and rose 234 feet above the plateau in 1906. Erosion began in the 1920s, and by 1961, the majestic giant had been reduced to only 132 feet in height. Studies on Grand Sable

Dunes indicate that both vehicle use and heavy foot traffic have led to increased erosion on grass- and herb-dominated portions of the dunes.

Interdunal swales are much less common than within the parabolic dunes. They also differ, as their water table is not determined by fluctuations in Great Lakes water levels, since they are perched far above the Great Lakes.

Beaches associated with perched dunes, lying beneath steep, eroding bluffs, are often much narrower than their parabolic dune counter-parts. During periods of high Great Lakes water levels, the lake's water can be directly eroding the bluffs. Even though the beaches can be quite narrow, piping plovers are known to nest on the beaches below the perched dunes, as at Sleeping Bear Dunes National Lakeshore.

Michael Penskar

Dunewort *(Botrychium campestre).*

Photo courtesy of The Nature Conservancy.

Clustered broom rape
*(Orobanche fasciculata).*

# ERODING BLUFF

**Steep eroding bluffs of till form the base from which the perched dunes rise.**

Swallow nests in the steep bluffs at Sleeping Bear Dunes National Lakeshore.

White camas (*Zigadenus glaucus*) growing on gravel lag.

For both the Grand Sable dunes on Lake Superior and the Sleeping Bear dunes on Lake Michigan, the bluffs are responsible for roughly 300 feet of elevation, while the perched dunes formed on the bluff are only half that height. The bluff faces of perched dunes are quite variable. At both Grand Sable and Sleeping Bear dunes, the bluff faces are very sandy, supporting dune plants, but fine-textured bands of clay or loam are also encountered. On some bluff faces fine-textured banding results in numerous seepages. Trees and shrubs establish on the bluff faces, but continual erosion of the bluff base results in widespread instability. Fallen trees are common on forested portions of the bluffs.

Both clustered broom rape and dunewort commonly grow on the eroding sandy bluffs at Sleeping Bear and Grand Sable dunes, and elsewhere. Pitcher's thistle also grows on both perched dunes and eroding bluffs. Colonies of bank swallows build nests into steep, cliff-like sections of bluff.

Harebell (*Campanula rotundifolia*) on gravel lag.

34

# Zonation of Dune & Swale

Harold Malde

Dwarf lake iris *(Iris lacustris)*.

Dune and swale complexes have four distinct, alternating zones. These are the **beach and foredune**, the **open interdunal swales**, the **forested dune ridges**, and the **forested swales**. Dune and swale complexes can be quite extensive; one of the largest is the 40 mile long Tolston Beach complex along the southern Lake Michigan shoreline in Indiana and Illinois.

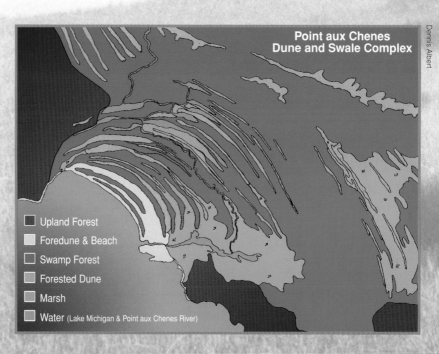

**Point aux Chenes Dune and Swale Complex**

Dennis Albert

- ☐ Upland Forest
- ☐ Foredune & Beach
- ☐ Swamp Forest
- ☐ Forested Dune
- ☐ Marsh
- ☐ Water (Lake Michigan & Point aux Chenes River)

# BEACH AND FOREDUNE

**The beach zone of dune and swale complexes and parabolic dune complexes share many processes, plants, and animals.**

In contrast, there is reduced sand availability on the foredunes of most dune and swale complexes. As a result, the dune and swale complexes never develop high parabolic dune features, although many contain small parabolic dunes. Typical linear beach ridges are only about 10 to 15 feet high, and about 30 to 100 feet wide.

Bare sand and herbaceous vegetation characterize the foredune, and often two or three additional beach ridges farther inland. As on the foredunes of parabolic dune complexes, marram grass is a major pioneering plant on the beach ridge closest to the shoreline. Shrub willows and sand cherry are also found here. But a major difference is seen on the second or third dune ridge farther from the shoreline, where marram grass is typically replaced by little bluestem, sand reed

grass, and low shrubs. The rapid decline of marram grass is due to the lack of annual sand deposition.

A recent study in northern Michigan demonstrated that within 200 years, on the fourth or fifth beach ridge from the shoreline, mixed pine forests had replaced the pioneering herbs and low shrubs of the shoreline. Similar patterns of vegetation succession are seen within most dune and swale complexes throughout northern Michigan. Where the dune features are lower and closer together, forest succession often occurs even more rapidly. The study also documented the gradual increase in organic material, moisture holding capacity, and nutrients on these 200-year-old dune soils.

Plants characteristic of the foredunes include creeping juniper, hoary puccoon, beach pea, wormwood, harebell, and common milkweed. Two Great Lakes endemic plants, Pitcher's thistle and Lake Huron tansy, occur on the open foredune, along with endemic Lake Huron locust.

Maureen Houghton

Foredune of dune and swale complex, Grass Bay on Lake Huron.

Another endemic plant, dwarf lake iris, is sometimes found on the shrubby, inland edge of the foredune, but it grows more vigorously on the sheltered inner ridges, where it forms dense carpets. It also persists, but does not flower, under closed forest canopies.

Lacking both adequate moisture and nutrients, the environment of the foredune is typically too extreme for the successful establishment of most upland tree species. As the shrub cover increases on ridges farther from shore, tree seedling establishment increases. In the north, white spruce, white pine, and red pine are common pioneering tree species, while cotton-wood, sassafras, and black oak are common pioneers in the south.

# OPEN INTERDUNAL SWALES

**The wetness of interdunal swales near the present lakeshore is determined by the Great Lakes water level.**

Shallow marly swale at Pointe aux Chenes dune and swale complex.

Dennis Albert

When water levels are high, several swales near the shoreline may be flooded. Swales farther from the shoreline can also be flooded, but these swales are typically flooded because of ground-water flow or seepage, not due to the lake's water level. Aquatic vegetation in the swales can be quite diverse, including aquatic or emergent grasses, sedges, herbs, and shrubs. In some complexes, there can be several open, sedge- or alder-dominated swales separated by forested ridges before the swales become consistently forested.

Aquatic insects are likewise abundant in these shallow, warm swales, along with leeches. Several populations of Hine's emerald dragon-flies, a federally endangered species, have recently been discovered within the marly swales near St. Ignace, on northern Lake Huron. Woodcock and common snipe commonly nest in the brushy swales, found where the open swales along the shoreline transition to forested swales.

The moist swale margins in the Straits of Mackinac, along the northern Lake Huron and Lake Michigan shoreline, are the primary habitat for a rare Great Lakes endemic plant, Houghton's goldenrod. During high Great Lakes water levels, Houghton's goldenrod

Butterwort *(Pinguicula vulgaris).*

Harold Malde

Houghton's goldenrod *(Solidago houghtonii).*

Susan Crispin

populations can decrease drastically and sometimes the plant may appear to be absent from a site. When water levels drop, the number of stems around a swale can increase dramatically, sometimes even occupying the entire bottom of moist swales. Houghton's golden-rod is occasionally found growing on the drier low dunes as well. The marly swales in the Straits area provide habitat for another showy rare plant, butterwort, which can be locally abundant along the moist margins or within the shallow waters of the swale. The size of butterwort populations can likewise fluctuate greatly in response to water level changes.

Pitcher-plant *(Sarracenia purpurea).*

Grass pink *(Calopogon tuberosus).*

Indian paintbrush *(Castilleja coccinea).*

Aerial photo of dune and swale complex, Grass Bay on Lake Huron.

# FORESTED DUNE RIDGES

**Succession on the dune and swale complexes is much more rapid than succession on the higher parabolic dunes, probably because of reduced sand accumulation in the dune and swale complex, thus allowing trees to establish without being buried.**

Shallow swale with aquatic plants, surrounded by conifer-dominated dunes, Gulliver, northern Lake Michigan.

Michael Penskar

Stabilization of the dunes by trees also results in relatively rapid soil development, with increased nutrient and moisture availability. This results in continual forest dominance, but replacement of pine forests by more mesic hardwood forest types appears to be very slow on dune and swale complexes, with conifers maintaining dominance for over 2,000 years. This is in strong contrast to many of the parabolic dune complexes, where hardwoods commonly dominate the more protected inner portions of the dunes. This continued dominance by pines, balsam fir, and early successional hardwoods, such as oaks, balsam poplar, and aspen, may result from regularly occurring wind storms. Both from the original surveyors' notes and from recent records, it is apparent that wind storms occur often near the shoreline, creating the large openings required for establishment of pines and other early successional tree species.

Large white and red pine on the ridges provide nest trees for bald eagles and ospreys, both of which feed either in the nearby lake or within the larger, open swales. During the summer months, ground-nesting northern harriers are encountered gliding gracefully above the open swales. Following large wind storms or major insect infestations, woodpeckers forage for insects and nest in standing, dead trees. Dwarf lake iris can form dense colonies growing beneath both open and closed-canopied conifer forests just back from the shoreline, but in this habitat it seldom flowers, except when the forest canopy is opened by a storm. Dwarf lake iris may persist on shady, forested ridges as much as a mile from shore.

Betty Cottrille

Black-backed woodpecker
(*Picoides arcticus*).

39

# FORESTED SWALES

**The establishment of forests within the swales is much more gradual than the establishment of forests on the ridges.**

While herbaceous vegetation often completely dominates only the 3 or 4 swales closest to shoreline, shrubs can be prevalent well into a complex. The shrubs are gradually replaced by swamp conifers as one proceeds further from the shoreline and as the swales become drier. In southern Michigan, common shrubs are buttonbush, Michigan holly, meadowsweet, and willows, while further north, speckled alder, sweet gale, mountain holly, and shrub-sized black spruce, northern white cedar, and tamarack become increasingly domi-nant. On Lake Superior, where organic materials decompose slowly because of the cold climate and where the soils are more acid, bog vegetation dominates these shrub swamps, including bog laurel, cranberries, bog rosemary, leatherleaf, and Labrador-tea.

Further from the shoreline, swamp conifers become the dominant vegetation. Groundwater-fed streams enter most of the dune and swale complexes. In the oxygen-rich waters near these streams, cedar, speckled alder, and hardwoods, such as paper birch and trembling aspen are abundant. In portions of the swales with restricted groundwater influence, black spruce, tamarack, and acid-tolerant shrubs are prevalent. Organic soils accumulate in most forested swales, but are seldom greater than three feet deep.

These dense conifer swamps provide extensive habitat for orchids and other plants requiring sheltered, cool, shady conditions. Among the rare plants known from these complexes is Lapland buttercup, a state threatened plant with a mostly boreal distribution, and round-leaved orchid, which also requires the cool protected habitat.

Faunal surveys of the dune and swale complex have been much less intensive than the plant surveys. Some rela-tively common users of this protected habitat are snow-shoe hare, common snipe, red fox, and whitetail deer. Whitetail deer utilize the thermal protection, as well as the abundant conifers for winter browse. Many dune and swale complexes are managed specifically for whitetail deer, with upland logging used to stimulate aspen growth for browse.

Studies of migratory song-birds along northern Lake Huron have demonstrated that coastal wetlands are an extremely important source of forage in the spring. During April and May, the coastal wetlands produce large numbers of midges at a time when there are few other food sources for migratory songbirds. The midges congregate on the needles of conifers, where warblers and other songbirds come to feast.

Gary Reese

Lapland buttercup *(Ranunculus lap-ponicus)* leaf on the right, with look alike goldthread *(Coptis trifolia)* leaf on the left.

# Zonation of Transverse Dunes Within Glacial Embayments

Betty Cottrille

Common snipe *(Gallinago gallinago)*.

There has been less detailed study of the dune fields within the glacial bays of the Upper Peninsula of Michigan than in any other dune type. Yet, even with this paucity of information, it is clear that these dune complexes are extremely important for maintaining Michigan's plant and animal diversity.

Dennis Albert

Red pine and white pine growing on low transverse dune ridges within extensive peatland, Barfield Lakes.

# DUNE FORESTS

**The original surveyors' notes from the mid-1800s described vast peatlands, often flooded by beaver dams, and broken by narrow white pine- and red pine-capped ridges.**

Low white-pine dominated transverse dunes within a shallow, sedge-dominated peatland, Barfield Lakes.

Dennis Albert

While white pine and red pine were common dominant trees, the surveyors also made reference to aspen and jack pine. In a few places, eastern hemlock grew with white pine, creating a majestic forest. A few examples of this forest type still remain, as at Swamp Lakes and portions of Barfield Lakes. Many of the ridges were at least partially burnt, as were the surrounding peatlands. Following settlement, winter logging was conducted on most of the bays, and fires often followed logging, sometimes resulting in the loss of pines and replacement by aspen. Plant surveys of the ridges have resulted in no rare plant finds. Bracken fern, blueberries, huckleberries, poverty grass, and Pennsylvania sedge are among the most common plants. It is also common to encounter bald-faced hornet nests, either in the loose sand or hanging from bracken fern, and in the twilight, one may be serenaded by bobolinks. The upland conifer forests remain important for birds like the pine siskin, Blackburnian warbler, and other songbirds requiring mature conifers for nesting.

Open, transverse dune ridge following heavy logging and slash fires.

Dennis Albert

American kestrel *(Falco sparverius).*

Janet Haas

**The peatlands surrounding the transverse dunes form an intriguing, diverse habitat. At the margins of many peatlands, there is typically a zone with groundwater influx, where a narrow band of northern white-cedar grows.**

Cedar quickly gives way to black spruce and tamarack, sometimes followed by stunted jack pine in the center of the peatland. Tree growth is slow because of the harsh, frost prone micro-climate and low nutrient availability. As a result of these extreme conditions, a 1-inch diameter tamarack may be 75 to 100 years old.

The swamp edges can be relatively diverse, but the center of the peatlands may have a depauperate flora of sphagnum mosses and a few species of sedge. Fire is no stranger to the peatlands, which can be quite dry during periods of drought. The original surveyors mapped large fires in several peatlands and thin bands of charcoal are common in the peat.

One of the most common animals within these large wetlands is the beaver. Small streams flowing across the wetlands will usually be blocked by a string of beaver dams. The habitat is important for other animals as well. During the spring and summer, the visitor can expect to be treated to the raucous, primitive calls of sandhill cranes, which feed on the abundant insects within the dune complex. Scattered small lakes within the peatland will occasionally harbor nesting loons or black terns. In the spring, sharp-tailed grouse can be seen engaging in their mating rituals on low rises within the wetlands, sometimes sending water spraying in all directions. Yellow rail, a secretive waterbird of bogs, also feeds and mates in these wetlands, using open, sedge- or grass-dominated habitat. Yellow-rail nests are found in shallow water within extensive stands of *Carex lasiocarpa*, a sedge, and populations of the rail increase following fires. American bittern is another

Labrador-tea *(Ledum groenlandicum)*.

Swamp-laurel *(Kalmia polifolia)*.

Moose *(Alces alces)* within large peatland and dune complex.

43

bird common to this complex of dunes and wetlands.

Merlins, a small falcon species, may be seen soaring over the wetlands in search of prey. The vast peatlands provide habitat for many of its prey species, including song birds, small mammals, dragonflies, and other large insects. Merlins nest in large trees along the wetland edge, utilizing abandoned nests of American crows or common ravens, or even building nests in witches' broom, a parasitic mistletoe growing on black spruce trees.

Songbirds nesting in the herbaceous and shrubby portions of these wetlands include LeConte's and Lincoln's sparrow. Dead standing trees, resulting from either insect infestations, beaver flooding, or fires, provide nesting cavities for rare black-backed woodpeckers. Black-backed woodpeckers become locally common following one of these disturbances, only to disappear and move to another disturbed site within just a few years, as nesting and foraging conditions decline.

The dune-peatland complexes also provide habitat for birds with a more northern, boreal distribution, including rusty blackbirds and palm warblers. Two other northern species, the white-winged and red crossbills, breed in mature stands of red pine on the narrow dune ridges.

These large peatlands are very important for small mammals like the snowshoe hare, and for large mammals like wolf, bear, and moose as well. With some of the lowest road densities in the state, bear and wolf are at home in these vast peatlands. Moose, recently reintroduced to the Upper Peninsula of Michigan, are also occasionally sighted within these peatlands.

Sandhill crane *(Grus canadensis)*.

Betty Cottrile

Typical peatland within complex of transverse dunes.

Harold Malde

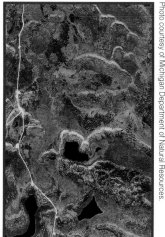

Color-infrared aerial photograph of glacial Lake Algonquin bay with transverse dune ridges, Ramsey-Lost Lakes.

Photo courtesy of Michigan Department of Natural Resources.

# Dune Threats

Both the ecological importance and recreational value of Michigan's sand dunes are well recognized. Unfortunately, there are also numerous threats to the dunes, ranging from degradation of the dunes by residential and recreational uses to complete elimination by mining. Among the threats identified along Michigan's coastal dunes, the following are among the most widespread or severe:

These condominiums result in the destruction of forested dune and provide easy shoreline access for ORVs.

Courtesy of Michigan Department of Environmental Quality.

- Exotic (introduced non-native) plants and animals.

- Off-road vehicles (ORVs)

- Pedestrian recreational overuse

- Residential development

- Sand mining and other industrial development

Todd Thompson

A highway and driveways cross this dune and swale complex, altering the hydrology.

45

**Exotic plants are a widespread threat to dunes.**

Photo courtesy of Gillette Nature Center.

Silver poplar *(Populus alba)*.

Walter Loope

Bladder campion *(Silene vulgaris)*.

One widespread exotic plant is spotted knapweed, which colonizes open habitat, often spreading along roadway corridors. It forms dense patches, resulting in stabilization of blowing sand, thus reducing habitat for rare species like Pitcher's thistle, which requires open, moving sand for its establishment and survival. Another plant that similarly stabilizes open sand dunes is baby's-breath. Both of these plants can be extremely difficult to eliminate once they have established, since their seed can persist in the sand, gradually germinating over

several years. Baby's-breath forms large taproots, up to several feet long, making manual removal difficult. Other herbaceous plants commonly invading the open dunes include bouncing bet, bladder campion, Canada bluegrass, autumn olive, scots pine, and smooth brome grass.

Still other plants have been purposely planted on the dunes by landowners to stabilize the dunes, especially black pine, Lombardy poplar, and white poplar. Black pine has begun to reproduce by seed on some of the state parks, causing a costly

management problem. Both Lombardy and white poplar reproduce vegetatively, producing open clones in the case of Lombardy poplar, and very dense clones for white poplar. Once trees have established on the dunes, the general public is sometimes opposed to their removal, even though these exotics may threaten populations of rare endemic dune plants.

To date, there are many less exotic animals than plants adversely impacting the dune environment. A recent arrival to the sandy shoreline is the zebra mussel. In bays of the Great Lakes, zebra mussel has been reported to cause chemical changes, resulting in increased levels of blue-green algae. While the effects on the sand beach have not yet been fully evaluated, the tremendous numbers of shells cannot help but alter some aspects of the dune ecosystem. These colorful shells often form windrows on the beach following storms, and can result in severe lacerations to the bare feet of beach-goers, especially children.

Another potential exotic threat is a beetle that has been introduced to control an exotic species of thistle. There are indications that the beetle may also feed on federally-threatened Pitcher's thistle.

Baby's-breath *(Gypsophila paniculata)*.

Lombardy poplar *(Populus nigra var. italica)*.

47

## Degradation: ORVs and Recreational Overuse

Earl Wolf

A visually pleasing boardwalk reduces the erosion at Hoffmaster State Park's Gillette Nature Center.

The effect of off-road vehicles and, in some cases, overly heavy pedestrian use have been well documented on the dunes. Probably the most dramatic example of off-road vehicle use is Silver Lake State Park, where extreme levels of use have eliminated almost all vegetation from a large portion of the dunes, creating conditions that we associate with the most barren, extreme desert environments. Studies done using aerial photos taken over a period of fifty years document the effect of off-road vehicles at both Sleeping Bear Dunes National Lakeshore and Grand Sable Dunes (Pictured Rocks National Lakeshore). The Great Sleeping Bear dune lost over half of its height from heavy vehicle use. Similarly, studies have shown that the amount of unvegetated sand increased dramatically at Grand Sable dunes as a result of vehicle and pedestrian uses.

Extreme pedestrian use has led to the development of boardwalks at some of the most heavily used parks. Boardwalks have been successful in reducing recreational erosion at several Michigan parks, including Hoffmaster and Warren Dunes State Parks.

Human degradation to the dunes takes many forms. Garbage and debris are widespread. Not only does refuse detract from the scenic beauty of the dunes, it can also be dangerous to wildlife. For example, shore birds can become entangled in plastic 6-pack holders or swallow the colorful remains of helium balloons and their attached ribbons. Helium balloons, purposefully or accidentally released along the shorelines from as far away as Wisconsin, are probably the most common debris found on the beach.

Dave Kenyon

Dune buggy, Silver Lake State Park.

Photo courtesy of Michigan State Parks.

The "dune climb" at Sleeping Bear Dunes National Lakeshore. Heavy pedestrian traffic has resulted in almost complete elimination of vegetation.

Dennis Albert

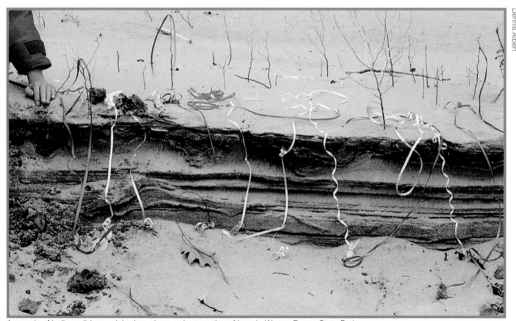

A sample of balloon ribbons picked up along a short section of beach, Warren Dunes State Park.

# RESIDENTIAL DEVELOPMENT

**There is probably no more attractive place to drink morning coffee than from one's dining room perched atop a coastal dune, yet, in terms of the coastal environment, homes built within the dunes cause numerous problems.**

Courtesy of Michigan Department of Environmental Quality.

Indiscriminate condominium and home development along the shoreline resulted in the passage of Michigan's sand dune regulations.

Every home in the dunes requires a driveway, and each driveway or road creates a corridor for exotic plants to enter the dunes. As soon as the home is built, the open sand which was only slightly objectionable when tracked onto the picnic blanket, becomes a major irritation, and a lawn is soon planted. Soon, blue grass, other lawn grasses, and accompanying weeds such as dandelion and smooth brome, become troublesome exotic pests on the open dune. When the sand begins burying the deck, swimming pool, or living room windows, the lawn is followed by a protective barrier of black pine or Lombardy poplar, providing yet another set of dune invaders.

Blowing and eroding sand limits appropriate locations for road and driveway construction. State and local zoning restrict driveway construction to slopes less than 33 percent, but widespread violation of these ordinances is common.

Courtesy of Michigan Department of Environmental Quality.

Erosion during high water periods can result in major property loss for those who built their homes too close to coastal bluffs.

Courtesy of Michigan Department of Environmental Quality.

This home along southern Lake Michigan was destroyed during the high-water period of the late 1980s.

Home construction on sand dunes can be a risky and an expensive venture. Sand dunes are by definition unstable. Increased activity, either in the form of human disturbance to dune plants or climatic disturbances, can result in destabilization of the dunes, resulting in rapid sand movement. It is not unusual during a study of coastal residential areas to encounter homes whose foundations have been eroded by bluff erosion or homes, swimming pools, or driveways buried by sand. Even the forested dunes can be a very risky place to build, as was demonstrated by a windstorm in late spring of 1998. The storm caused extensive property damage within the dunes near Muskegon, when thousands of uprooted trees destroyed roofs, blocked roads and driveways, and damaged cars. There have been recent, severe storms along the northern Lake Michigan shoreline as well, with large white pine and red pine especially prone to wind damage.

Residential development can also seriously degrade linear dune (dune and swale) complexes. To access the shoreline, where most residential development occurs within these complexes, a series of driveways are built crossing the entire series of dunes and swales, resulting in severe altera-tion of the drainage conditions within the swales. Swales that previously provided abundant wildlife habitat, become a series of cat-tail- or purple loosestrife-choked, stagnant ponds. Homes built close to the shoreline often stabilize foredunes. Habitat for threatened endemic species like the Lake Huron locust, Pitcher's thistle, dwarf lake iris, Houghton's goldenrod, and Lake Huron tansy is destroyed or degraded by alteration of habitat or the introduction of exotic plants like baby's-breath, spotted knapweed, or Kentucky blue grass.

Courtesy of Michigan Department of Environmental Quality.

Todd Thompson

A highway and driveways cross this dune and swale complex, altering the hydrology.

Creative fantasy becomes coastal debris when Great Lakes waters rise.

51

# OTHER DEVELOPMENT

**In the recent past, coastal dunes were an extremely important source of sand for industrial uses.**

R. Torresen

Agricultural and residential development behind the dunes reduces the value for wildlife species like the eastern box turtle or nesting songbirds. Rosy Mound, southern Lake Michigan.

One of the most significant industrial uses of dune sand was for creating foundry casting molds. Michigan's automobile industry relied extensively on sands from the Lake Michigan shoreline until the 1980s when public pressure forced the industry to find other sources of sand. As our need for sand continues, inland sands have proved to be an economically viable sand source, somewhat reducing the pressure on our coastal dunes. Other industrial uses of sand include road construction and winter road treatment, construction fill, and glass production. In the Muskegon area, glass manufacturing resulted in the complete removal of some coastal dunes. Sand mining continues in many Great Lakes coastal areas.

Presently, one of the first dune restoration projects is underway at a sand mining site adjacent to Grand Mere State Park, near Bridgeman. Other sand mining sites have been stabilized rather than restored, using the ponds or lakes created during mining operations as focal points for golf courses or condominium complexes.

Golf course development linked to luxury homes or condominiums often destroys the ecological integrity of sand dunes. Most golf courses replace dune grasses with turf grasses, both eliminating habitat for native dune species and introducing exotic plants into the dune ecosystem. Forest vegetation is often removed, altering the micro-climate, which can result in dune destabilization. In recent years there was a dramatic example of the disastrous effects of such an alteration of dune habitat, when forest removal, hydrological manipulations, and turf-grass establishment along a dune bluff resulted in tremendous levels of erosion and ultimately also the loss of the scenic green.

Donald Petersen

Sand mining along southern Lake Michigan.

# SAND DUNE REGULATIONS

**The Sand Dunes Protection and Management Program, as part of the Natural Resources and Environmental Protection Act, 1994 PA 451, is in place to protect dunes from residential, commercial, and industrial uses.**

Sand dune regulations were enacted by the state of Michigan to protect dunes that were not publicly owned from indiscriminate development. Seventy thousand acres of shoreline along Lake Michigan and Lake Superior were designated and mapped by the Michigan Department of Natural Resources as "critical dune areas." These areas are mapped in the Atlas of Critical Dunes (1989). Critical dune areas include, but are not limited to, all barrier dunes along the shoreline, both open and forested dunes. Parabolic, perched, and linear dunes are included within the "critical dune areas" designation.

Not only do the sand dune regulations identify the dunes being regulated, they also restrict certain activities within the dunes. Construction of structures, roads, and driveways is now restricted on steep slopes (over 33%) to reduce dune destabilization. Before the regulations were enacted, construction of homes on steep slopes and unstable, open dunes was common.

Home or condominium construction along the shoreline periodically results in the loss of structures, especially during high-water periods, when dune and shoreline bluff erosion can be quite rapid. The loss of structures can be quite costly to the landowner. Attempts to save homes by building jetties or other shoreline structures result in the disruption of along-shore currents and sediment transport, sometimes negatively affecting nearby properties.

While current regulations have resulted in reduced degradation of Michigan's sand dunes, development pressure remains high. The density of development in many dune areas has been very high, resulting in loss of many ecological values. While regulations provide some protection for the dunes, development of the adjacent, flatter sand lake plain has been even more intense. In many places, there is no natural vegetation to provide a buffer between the dunes and the adjacent urban, industrial, or agricultural lands. This buffering is critical for the survival of many of the plants and animals utilizing the dunes. Lack of buffer is even seen at some of the state's premier parks, where homes, golf courses, and even sand mining operations are within easy view of the parks. Unfortunately, sand dune regulations have not eliminated this intensive development, as demonstrated at Hoffmaster State Park, where new homes on the foredune extend to within a few hundred feet of the park. Such development visually detracts from the park's beauty and can provide easy access for off-road vehicle trespass into the parks.

# DUNES IN MICHIGAN - LOWER PENINSULA

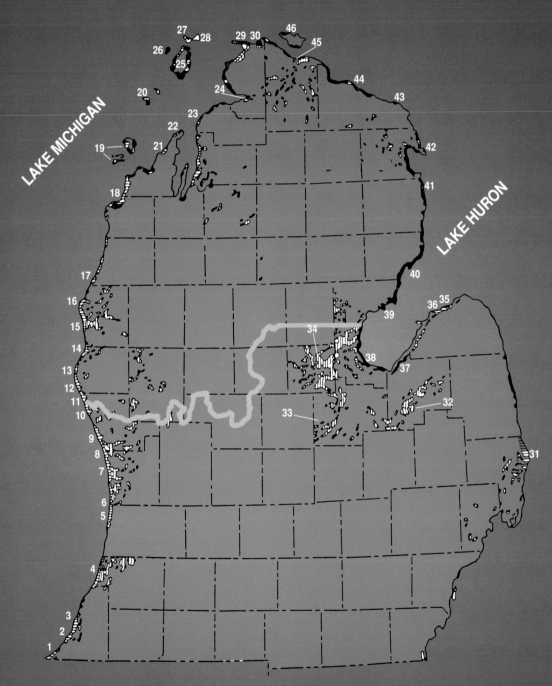

*Remember that dune habitats are fragile and easily damaged by human activities.*
*Please use boardwalks or special viewing platforms when available.*

# SAND DUNES ON PUBLICLY-OWNED LANDS OR PRIVATE NATURE PRESERVES IN LOWER MICHIGAN

| | Dune & Swale Complex (Linear Dunes) | | Parabolic Dunes | | Perched Dunes | | Transverse Dunes | |
|---|---|---|---|---|---|---|---|---|
| | South | North | South | North | South | North | South | North |
| 1. Grand Beach | X | | | | | | | |
| 2. **Warren Dunes State Park** | | | X | | | | | |
| 3. Grand Mere State Park | | | X | | | | | |
| 4. Van Buren State Park | | | X | | | | | |
| 5. **Saugatuck State Park** | | | X | | | | | |
| 6. Holland State Park | | | X | | | | | |
| 7. Rosy Mound State Park | | | X | | | | | |
| 8. **Hoffmaster State Park*** | | | X | | | | | |
| 9. **Muskegon State Park*** | | | X | | | | | |
| 10. Duck Lake State Park | | | X | | | | | |
| 11. Meinert County Park | | | | X | | | | |
| 12. Benona Township Park | | | | X | | | | |
| 13. Silver Lake State Park | | | | X | | | | |
| 14. Charles Mears State Park | | | | X | | | | |
| 15. **Ludington State Park*** | | | | X | | | | |
| 16. **Nordhouse Dunes Wilderness Area (USFS)** | | | | X | | | | |
| 17. Orchard Beach State Park | | | | X | | | | |
| 18. **Sleeping Bear Dunes National Lakeshore*** | | X | | | | X | | |
| 19. **Manitou Islands; Sleeping Bear Dunes N. L.** | | X | | | | X | | |
| 20. South Fox Island | | | | | | X | | |
| 21. Leland Municipal Beach | | X | X | | | | | |
| 22. Leelanau State Park | | | | X | | | | |
| 23. Fishermans Island State Park | | X | | | | | | |
| 24. Petoskey State Park | | | | X | | | | |
| 25. Beaver Island (MI DNR) | | X | | X | | | | |
| 26. High Island (State Wildlife Research Area) | | | | | | X | | |
| 27. Garden Island (MI DNR) | | X | | | | | | |
| 28. Hog Island (MI DNR) | | X | | | | | | |
| 29. **Wilderness State Park** | | X | | X | | | | |
| 30. Trail's End Bay (Twp. Park) | | X | | | | | | |
| 31. Lakeport State Park | X | | | | | | | |
| 32. Vassar and Tuscola State Game Areas | | | | | | | X | |
| 33. Gratiot-Saginaw State Game Area | | | | | | | X | |
| 34. Au Sable State Forest | | | | | | | X | |
| 35. **Port Crescent State Park** | X | | | | | | | |
| 36. **Albert E. Sleeper State Park** | X | | | | | | | |
| 37. Fish Point Wildlife Area | X | | | | | | | |
| 38. Tobico State Game Area | X | | | | | | | |
| 39. Pointe Au Gres (MI DNR) | | X | | | | | | |
| 40. Tawas Point State Park | | X | | | | | | |
| 41. **Negwegon State Park** | | X | | | | | | |
| 42. Misery Bay (MI DNR) | | X | | | | | | |
| 43. Thompsons Harbor State Park | | X | | | | | | |
| 44. Hoeft State Park | | X | | | | | | |
| 45. **Cheboygan State Park** | | X | | | | | | |
| 46. Point Catosh, Bois Blanc Island (MI DNR) | | X | | | | | | |

**Codes:**
**Bolded sites** are excellent examples of dune types with easy access.

\* **Interpretative Services Available.**

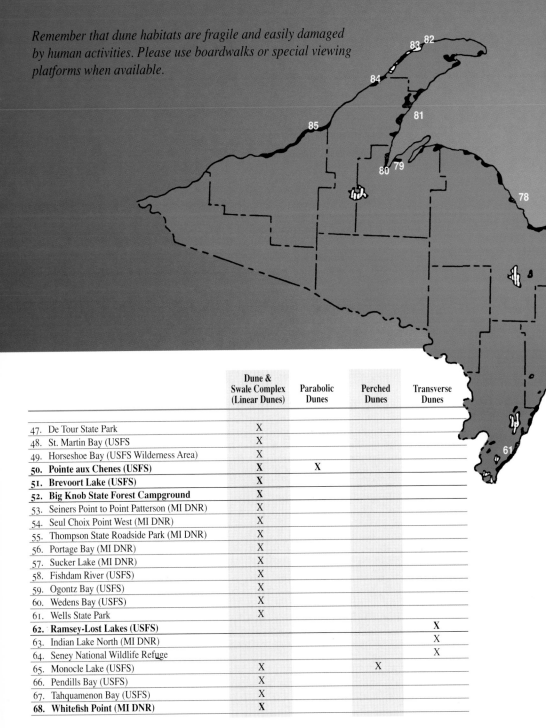

# DUNES IN MICHIGAN - UPPER PENINSULA

Remember that dune habitats are fragile and easily damaged
by human activities. Please use boardwalks or special viewing
platforms when available.

| | Dune & Swale Complex (Linear Dunes) | Parabolic Dunes | Perched Dunes | Transverse Dunes |
|---|---|---|---|---|
| 47. De Tour State Park | X | | | |
| 48. St. Martin Bay (USFS | X | | | |
| 49. Horseshoe Bay (USFS Wilderness Area) | X | | | |
| **50. Pointe aux Chenes (USFS)** | X | X | | |
| **51. Brevoort Lake (USFS)** | X | | | |
| **52. Big Knob State Forest Campground** | X | | | |
| 53. Seiners Point to Point Patterson (MI DNR) | X | | | |
| 54. Seul Choix Point West (MI DNR) | X | | | |
| 55. Thompson State Roadside Park (MI DNR) | X | | | |
| 56. Portage Bay (MI DNR) | X | | | |
| 57. Sucker Lake (MI DNR) | X | | | |
| 58. Fishdam River (USFS) | X | | | |
| 59. Ogontz Bay (USFS) | X | | | |
| 60. Wedens Bay (USFS) | X | | | |
| 61. Wells State Park | X | | | |
| **62. Ramsey-Lost Lakes (USFS)** | | | | X |
| 63. Indian Lake North (MI DNR) | | | | X |
| 64. Seney National Wildlife Refuge | | | | X |
| 65. Monocle Lake (USFS) | X | | X | |
| 66. Pendills Bay (USFS) | X | | | |
| 67. Tahquamenon Bay (USFS) | X | | | |
| **68. Whitefish Point (MI DNR)** | X | | | |

# SAND DUNES ON PUBLICLY-OWNED LANDS OR PRIVATE NATURE PRESERVES IN UPPER MICHIGAN

| | Dune & Swale Complex (Linear Dunes) | Parabolic Dunes | Perched Dunes | Transverse Dunes |
|---|:---:|:---:|:---:|:---:|
| **69. Tahquamenon Falls State Park** | X | | | X |
| 70. Swamp Lakes (Preserve: The Nature Conservancy) | | | | X |
| 71. McMahon Lake(MI DNR and Nature Conservancy Preserve) | | | | X |
| 72. Two-Hearted River Mouth | X | X | | |
| 73. Muskallonge Lake to Grand Marais (MI DNR) | X | | | X |
| **74. Grand Sable Dunes (Pictured Rocks Nat'l. Lakeshore)*** | | | | X |
| 75. 12-Mile Beach (Pictured Rocks Nat'l. Lakeshore) | X | | | |
| **76. Grand Island (USFS Federal Recreation Area)** | X | | | |
| 77. Au Train Bay (USFS and MI State Roadside Park) | X | | | |
| 78. Little Presque Isle (MI DNR) | X | | | |
| 79. Pequaming (MI DNR and TWP. Park) | X | | | |
| 80. Baraga State Park | X | | | |
| 81. Traverse Bay County Park | X | | | |
| 82. Eagle Harbor (Michigan Nature Association Preserve) | X | X | | |
| 83. Great Sand Bay (Michigan Nature Association Preserve) | | | X | |
| 84. F. J. McLain State Park | X | | | |
| 85. Sleeping Bay and Misery Bay (MI DNR) | X | | | |

**Codes:**
**Bolded sites** are excellent examples of dune types with easy access.

**\* Interpretative Services Available.**

# REFERENCED SPECIES: COMMON AND LATIN

## PLANTS OF THE DUNES AND ASSOCIATED WETLANDS REFERENCED IN REPORT

| Common Name | Latin Name | Common Name | Latin Name | Common Name | Latin Name |
|---|---|---|---|---|---|
| American beech | *Fagus grandifolia* | Eastern hemlock | *Tsuga canadensis* | Red oak | *Quercus rubra* |
| Aspen, big-toothed | *Populus grandidentata* | Grass-pink | *Calopogon tuberosus* | Red anemone | *Anemone multifida* |
| Aspen, trembling | *Populus tremuloides* | Ground juniper | *Juniperus horizontalis* | Red pine | *Pinus resinosa* |
| Baby's-breath | *Gypsophila paniculata* | Hairy puccoon | *Lithospermum caroliniense* | Round-leaved orchid | *Amerorchis rotundifolia* |
| Balsam fir | *Abies balsamea* | Harebell | *Campanula rotundifolia* | Rue anemone | *Anemonella thallictroides* |
| Balsam poplar | *Populus balsamifera* | Hepatica | *Hepatica americana,* | Rush | *Juncus sp.* |
| Basswood | *Tilia americana* | | *H. acutiloba* | Sand cherry | *Prunus pumila* |
| Beach pea | *Lathyrus japonicus* | Houghton's goldenrod | *Solidago houghtonii* | Sand reed grass | *Calamovilfa longifolia* |
| Bearberry | *Arctostaphylos uva-ursi* | Huckleberry | *Gaylussacia baccata* | Sassafras | *Sassafras albidum* |
| Bishop's cap | *Mitella diphylla* | Indian paintbrush | *Castilleja coccinea* | Sea rocket | *Cakile edentula* |
| Black pine | *Pinus nigra* | Jack pine | *Pinus banksiana* | Seaside spurge | *Euphorbia polygonifolia* |
| Black spruce | *Picea mariana* | Jack-in-the-pulpit | *Arisaema triphyllum* | Sedge | *Carex sp.* |
| Black oak | *Quercus velutina* | Labrador tea | *Ledum groenlandicum* | Silver poplar | *Populus alba* |
| Bloodroot | *Sanguinaria canadensis* | Lake Huron tansy | *Tanacetum huronense* | Solomon's seal | *Polygonum pubescens* |
| Blueberry | *Vaccinium spp.* | Lapland buttercup | *Ranunculus lapponicus* | Speckled alder | *Alnus rugosa* |
| Bog rosemary | *Andromeda glaucophylla* | Leatherleaf | *Chamaedaphne calyculata* | Spotted knapweed | *Centauria maculata* |
| Bog laurel | *Kalmia polifolia* | Little bluestem | *Schizachyrium scoparius* | Spring-beauty | *Claytonia virginica* |
| Bracken fern | *Pteridium aquilinum* | Marram grass | *Ammophila breviligulata* | Squirrel-corn | *Dicentra canadensis* |
| Bugseed | *Corispermum hyssopifolium* | Marsh-marigold | *Caltha palustris* | Star flower | *Trientalis borealis* |
| Bulrush | *Scirpus sp.* | May-apple | *Podophyllum peltatum* | Sugar maple | *Acer saccharum* |
| Bunchberry | *Cornus canadensis* | Meadowsweet | *Spiraea alba* | Swamp-laurel | *Kalmia polifolia* |
| Butterfly weed | *Asclepias tuberosa* | Michigan holly | *Ilex verticillata* | Sweet gale | *Myrica gale* |
| Butterwort | *Pinguicula vulgaris* | Moonwort | *Botrychium mormo,* | Tamarack | *Larix laricina* |
| Clustered broom rape | *Orobanche fasciculata* | | *B. acuminatum,* | Trout lily | *Erythronium americanum* |
| Common juniper | *Juniperus communis* | | *B. hesperium* | Virginia meadow-beauty | *Rhexia virginia* |
| Common milkweed | *Asclepias syriaca* | Mountain holly | *Nemopanthus mucronata* | White camas | *Zigadenus glaucus* |
| Common trillium | *Trillium grandiflorum* | Nodding trillium | *Trillium cernuum* | White spruce | *Picea glauca* |
| Cottonwood (Eastern) | *Populus deltoides* | Osier dogwood | *Cornus stolonifera* | White pine | *Pinus strobus* |
| Cranberry | *Vaccinium oxycoccos,* | Pennsylvania sedge | *Carex pensylvanica* | Willow | *Salix sp.* |
| | *V. macrocarpon* | Pitcher-plant | *Sarracenia purpurea* | Winged Pigweed | *Cycloloma atriplicifolia* |
| Creeping juniper | *Juniperus horizontalis* | Pitcher's thistle | *Cirsium pitcheri* | Wormwood | *Artemisia campestris* |
| Dunewort | *Botrychium campestre* | Poison ivy | *Toxicodendron radicans* | Yellow violet | *Viola pubescens* |
| Dutchman's-breeches | *Dicentra cucullaria* | Poverty grass | *Danthonia spicata* | | |
| Dwarf lake iris | *Iris lacustris* | Prairie phlox | *Phlox pilosa* | | |
| Eastern white-cedar | *Thuja occidentalis* | | | | |

## ANIMALS OF THE DUNES AND ASSOCIATED WETLANDS REFERENCED IN REPORT

| Common Name | Latin Name | Common Name | Latin Name | Common Name | Latin Name |
|---|---|---|---|---|---|
| **Amphibians** | | **Birds, continued** | | **Insects – Dragonflies** | |
| Blanchard's cricket frog | *Acris crepitans blanchardi* | Ovenbird | *Seiurus aurocapillus* | Hine's emerald | *Somatochlora hineana* |
| Blue-spotted salamander | *Ambystoma laterale* | Palm warbler | *Dendroica palmarum* | **Insects – Grasshoppers and Crickets** | |
| Fowler's toad | *Bufo fowleri* | Pine siskin | *Carduelis pinus* | Lake Huron locust | *Trimerotropis huroniana* |
| Northern leopard frog | *Rana pipiens* | Piping plover | *Charadrius melodus* | **Insects – Wasps and Hornets** | |
| Red-backed salamander | *Plethodon cinereus* | Prairie warbler | *Dendroica discolor* | Bald-faced hornet | *Vespula maculata* |
| Spring peeper | *Pseudacris crucifer crucifer* | Red crossbill | *Loxia curvirostra* | Spider (digger) wasps | *Pompilidae* (Family) |
| **Arthropods - Spiders** | | Red-eyed vireo | *Vireo olivaceus* | **Insects – Other** | |
| Wolf spider | *Lycosidae* (Family) | Red-shouldered hawk | *Buteo lineatus* | Antlion | *Myrmeleontidae* (Family) |
| **Birds** | | Rusty blackbird | *Euphagus carolinus* | **Mammals** | |
| American bittern | *Botaurus lentiginosus* | Sanderling | *Calidris alba* | Beaver | *Castor canadensis* |
| American kestrel | *Falco sparverius* | Sandhill crane | *Grus canadensis* | Black bear | *Ursus americanus* |
| Bald eagle | *Haliaeetus leucocephalus* | Scarlet tanager | *Piranga olivacea* | Mink | *Mustela vison* |
| Black tern | *Chlidonias niger* | Sharp-shinned hawk | *Accipiter striatus* | Moose | *Alces alces* |
| Black-and-white warbler | *Mniotilta varia* | Sharp-tailed grouse | *Tympanuchus phasianellus* | Muskrat | *Ondatra zibethicus* |
| Black-backed woodpecker | *Picoides arcticus* | Song sparrow | *Melospiza melodia* | Snowshoe hare | *Lepus americanus* |
| Black-billed cuckoo | *Coccyzus erythropthalmus* | Spotted sandpiper | *Actitis macularia* | White-tailed deer | *Odocoileus virginianus* |
| Black-throated green warbler | *Dendroica virens* | White-winged crossbill | *Loxia leucoptera* | Wolf | *Canis lupus* |
| Blackburnian warbler | *Dendroica fusca* | Wood thrush | *Hylocichla mustelina* | **Reptiles** | |
| Bluebird, eastern | *Sialia sialis* | Woodcock, American | *Scolopax minor* | Black rat snake | *Elaphe obsoleta obsoleta* |
| Broad-winged hawk | *Buteo platypterus* | Yellow rail | *Coturnicops noveboracensis* | Blue racer (snake) | *Coluber constrictor* |
| Common loon | *Gavia immer* | Yellow-billed cuckoo | *Coccyzus americanus* | Eastern hognose snake | *Heterodon platirhinos* |
| Common snipe | *Gallinago gallinago* | **Insects - Beetles** | | Eastern box turtle | *Terrapene carolina* |
| Cooper's hawk | *Accipiter cooperii* | Ladybugs | *Coccinellidae* (Family) | | *carolina* |
| Goshawk | *Circus cyaneus* | Tiger beetles | *Cicindela scutellaris* | Garter snake | *Thamnophis spp.* |
| LeConte's sparrow | *Ammodramus leconteii* | **Insects – Butterflies and moths** | | Spotted turtle | *Clemmys guttata* |
| Lincoln's sparrow | *Melospiza lincolnii* | Dune cutworm | *Euxoa aurulenta* | | |
| Merlin | *Falco columbarius* | Monarch butterfly | *Danaus plexippus* | | |
| | | Tiger Swallowtail | *Papilio glaucus* | | |
| | | White admiral | *Limenitis arthemus* | | |

# EXOTIC PLANTS TABLE

## COMMON EXOTIC PLANTS OF THE SAND DUNES

| Common Name | Latin Name |
| --- | --- |
| Autumn olive | *Elaeagnus umbellata* |
| Baby's-breath | *Gypsophila paniculata* |
| Black locust | *Robinia pseudoacacia* |
| Bladder campion | *Silene vulgaris* |
| Bouncing bet | *Saponaria officinalis* |
| Bull thistle | *Cirsium vulgare* |
| Canada bluegrass | *Poa compressa* |
| Chicory | *Cichorium intybus* |
| Clover | *Trifolium spp.* |
| Dandelion | *Taraxacum officinale* |
| Goat's beard | *Tragopogon sp.* |
| Hawkweed | *Hieracium caespitosum* |
| Heal-all | *Prunella vulgaris* |
| Honeysuckle | *Lonicera sp.* |
| Lombardy popular | *Populus nigra var. italica* |
| Mossy stonecap | *Sedum acre* |
| Mullein | *Verbascum thapsus* |
| Ox-eye daisy | *Chrysanthemum leucanthemum* |
| Quack grass | *Agropyron repens* |
| Redtop | *Agrostis gigantea* |
| Scots pine | *Pinus sylvestris* |
| Smooth brome | *Bromus inermus* |
| Spotted knapweed | *Centaurea maculosa* |
| Tall fescue | *Festuca arundinacea* |
| Timothy | *Phleum pratense* |
| White sweet clover | *Melilotus alba* |
| White poplar | *Populus alba* |
| Wild carrot | *Daucus carota* |
| Yarrow | *Achillea millefolium* |

# RARE PLANTS & ANIMALS

## RARE PLANTS OF THE DUNES AND ASSOCIATED WETLANDS

| Common Name | Latin Name | Dune & Swale Complex (Linear Dunes) | | Parabolic Dunes | | Perched Dunes | | Transverse Dunes | |
|---|---|---|---|---|---|---|---|---|---|
| | | South | North | South | North | South | North | South | North |
| Alga pondweed | *Potamogeton confervoides* | | | | | | | | X |
| American dune wild-rye | *Elymus mollis* | | S | S | | S | | | |
| American shore-grass | *Littorella uniflora* var. *americana* | | S | | | | | | X |
| Auricled twayblade | *Listera auriculata* | | S | | | | | | |
| Bald-rush | *Psilocarya scirpoides* | | | M | | | | | |
| Beauty sedge | *Carex concinna* | | M,H | | | | | | |
| Black-fruited spike-rush | *Eleocharis melanocarpa* | | | M | | | | | |
| Blue wild-rye | *Elymus glaucus* | | | | | | S | | |
| Butterwort | *Pinguicula vulgaris* | | **M,H** | | | | | | |
| Calypso | *Calypso bulbosa* | | **M,H,S** | | M,S | | M,S | | |
| Canada rice-grass | *Oryzopsis canadensis* | | | | | | | | X |
| Carey's smartweed | *Polygonum careyi* | | | M | | | | | |
| Climbing fumitory | *Adlumia fungosa* | H | | M | M | | | | |
| Cross-leaved milkwort | *Polygala cruciata* | | | M | | | | | |
| Douglas's hawthorn | *Crataegus douglasii* | | S | | | | S | | |
| Downy oat-grass | *risetum spicatum* | | | | | | S | | |
| Dunewort or prairie moonwort | *Botrychium campestre* | | | | | | M,S | | |
| Dwarf lake iris | *ris lacustris* | | **H,M** | | M | | | | |
| Dwarf-bulrush | *Hemicarpha micrantha* | | | | M | | | | |
| English sundew | *Drosera anglica* | | | | | | | | X |
| Fascicled broom-rape | *Orobanche fasciculata* | | | | **M** | | **M** | | |
| Frost grape | *Vitis vulpina* | | | M | | | | | |
| Ginseng | *Panax quinquefolius* | | | **M** | M | | | | |
| Hill's thistle | *Cirsium hillii* | | | M | | | | | |
| Houghton's goldenrod | *Solidago houghtonii* | | **M,H** | | M | | | | |
| Lake Huron tansy | *Tanacetum huronense* | | **M,H,S** | | M,S | | M,S | | |
| Lapland buttercup | *Ranunculus lapponicus* | | M | | | | | | |
| Meadow-beauty | *Rhexia virginica* | | | M | | | | | |
| Moonwort, acute-leaved | *Botrychium acuminatum* | | | | | | S | | |
| Moonwort, goblin | *Botrychium mormo* | | | | | | S | | |
| Moonwort, western | *Botrychium hesperium* | | | | | | S | | |
| Northern appressed clubmoss | *Lycopodiella subappressa* | | | M | | | | | X |
| Orange grass | *Hypericum gentianoides* | | | M | | | | | |
| Paniced hawkweed | *Hieracium paniculatum* | | | M | | | | | |
| Paniced screw-stem | *Bartonia paniculata* | | | | | | | | X |
| Pinedrops | *Pterospora andromedea* | | **H,M,S** | | S | | | | |
| Pitcher's thistle | *Cirsium pitcheri* | H,M | H,M | H,M | H,M | M | M,S | | |
| Pumpelly's brome grass | *Bromus pumpellianus* | | **M,S** | | **M,S** | | **M,S** | | |
| Ram's head lady's-slipper | *Cypripedium arietinum* | | H | | **M** | S | | | |
| Rose or swamp mallow | *Hibiscus moscheutos* | | | M | | | | | |
| Rose pink | *Sabatia angularis* | | | M | | | | | |
| Round-fruited loosestrife | *Ludwigia sphaerocarpa* | | | M | | | | | |
| Sand grass | *Triplasis purpurea* | | | M | | | | | |
| Satiny willow | *Salix pellita* | | S | | | | | | |
| Sedge | *Carex platyphylla* | | | M | | | | | |
| Sedge | *Carex seorsa* | | | M | | | | | |
| Showy orchis | *Galearis spectabilis* | | | M | | | | | |
| Spotted pondweed | *Potamogeton pulcher* | | | M | | | | | |
| Stitchwort | *Stellaria longipes* | | M,S | | M,S | | M,S | | |
| Tall beak-rush | *Rhynchospora macrostachya* | | | M | | | | | |
| Tall green milkweed | *Asclepias hirtella* | H | | | | | | | |
| Tuberous Indian plantain | *Cacalia plantaginea* | H | H | | | | | | |
| Whorled mountain mint | *Pycnanthemum verticillatum* | | | M | | | | | |
| Wiegand's sedge | *Carex wiegandii* | | | M | | | | | X |
| Wild bean | *Strophostyles helvula* | | | M | | | | | |
| Wild oatgrass | *Danthonia intermedia* | | M | | | | | | |
| Yellow ladies'-tresses | *Spiranthes ochroleuca* | | | M | | | | | |
| Zigzag bladderwort | *Utricularia subulata* | | | M | | | | | |

Codes: H = Lake Huron, M = Lake Michigan, S = Lake Superior,
X = inland dunes (associated with glacial lake beds, often several miles inland from present Great Lake shorelines)

**Bold** indicates strong association between dune community and the plant or animal species.

## RARE ANIMALS OF THE DUNES AND ASSOCIATED WETLANDS

| Common Name | Latin Name | Dune & Swale Complex (Linear Dunes) | | Parabolic Dunes | | Perched Dunes | | Transverse Dunes | |
|---|---|---|---|---|---|---|---|---|---|
| | | South | North | South | North | South | North | South | North |
| **AMPHIBIANS** | | | | | | | | | |
| Blanchard's cricket frog | Acris crepitans blanchardi | | | M | | | | | |
| **BIRDS** | | | | | | | | | |
| Bald eagle | Haliaeetus leucocephalus | H | M,H,S | M,S | | M,S | | X | |
| Black tern | Chlidonias niger | | S | | | | | | |
| Black-backed woodpecker | Picoides arcticus | | M,H,S | | | | | X | |
| Common loon | Gavia immer | | | | | | | X | |
| Merlin | Falco columbarius | | | | | | S | X | |
| Osprey | Pandion haliaetus | | S,M,H | | | | | X | |
| Peregrine falcon | Falco peregrinus | | | | | | M | | |
| Piping plover | Charadrius melodus | | M,H,S | M | M | M,S | | | |
| Tern, caspian | Sterna caspia | | | | M | M | | | |
| Tern, common | Sterna hirundo | H | H | | M | M | | | |
| Tern, Forester's | Sterna foresteri | H | | | | | | | |
| Warbler, cerulean | Dendroica cerulea | | | | M | | | | |
| Warbler, prairie | Dendroica discolor | | | M | M | M | | | |
| **INSECTS - BUTTERFLIES AND MOTHS** | | | | | | | | | |
| Aweme borer | Papaipema aweme | | M | | | | | | |
| Dune cutworm | Euxoa aurulenta | | | M | | | | | |
| Frigga fritillary | Boloria frigga | | S | | | | | X | |
| Grizzled skipper | Pyrgus centaureae wyandot | | **H** | | | | | | |
| **INSECTS - CICADAS AND HOPPERS** | | | | | | | | | |
| Spittlebug | Philaenarcys killa | | M | | | | | | |
| **INSECTS - DRAGONFLIES** | | | | | | | | | |
| Hine's emerald | Somatochlora hineana | | **H,M** | | | | | | |
| Incurvate emerald | Somatochlora incurvata | | S | | | | | X | |
| **INSECTS - GRASSHOPPERS AND CRICKETS** | | | | | | | | | |
| Lake Huron locust | Trimerotropis huroniana | H | **M,H,S** | M,H,S | | M,S | | | |
| Pinetree cricket | Oecanthus pini | | | M | | | | | |
| **MAMMALS** | | | | | | | | | |
| Moose | Alces alces | | | | | | | X | |
| Vole, prairie | Microtus ochrogaster | | | M | | | | | |
| Vole, woodland | Microtus pinetorum | | | M | | | | | |
| **REPTILES** | | | | | | | | | |
| Eastern fox snake | Elaphe vulpina gloydi | | H | | | | | | |
| Turtle, eastern box | Terrapene carolina carolina | | | M | M | | | | |
| Turtle, spotted | Clemmys guttata | | | M | | | | | |
| Turtle, wood | Clemmys insculpta | | | | M | | | | |

Codes: H = Lake Huron, M = Lake Michigan, S = Lake Superior,
X = inland dunes (associated with glacial lake beds, often several miles inland from present Great Lake shorelines)

**Bold** indicates strong association between dune community and the plant or animal species.

## RARE PLANT COMMUNITIES ASSOCIATED WITH DUNES AND DUNE WETLANDS

| Plant Community | Dune & Swale Complex (Linear Dunes) | | Parabolic Dunes | | Perched Dunes | | Transverse Dunes | |
|---|---|---|---|---|---|---|---|---|
| | South | North | South | North | South | North | South | North |
| Coastal plain marsh | M | | | | | | | |
| Lakeplain oak opening | H | | | | | | | |
| Lakeplain wet prairie | H | | | | | | | |
| Lakeplain wet-mesic prairie | H | | | | | | | |

Codes: H = Lake Huron, M = Lake Michigan, S = Lake Superior,
X = inland dunes (associated with glacial lake beds, often several miles inland from present Great Lake shorelines)

**Bold** indicates strong association between dune community and the plant or animal species.

# RECOMMENDED DUNE LITERATURE

*Atlas of Critical Dunes.*
Michigan Department of Natural
Resources Land and Water
Management Division. 1989.
(now in Michigan DEQ)

*A Field Guide to Great Lakes
Coastal Plants.* Walter J. Hoag-
man. Michigan Sea Grant Extension. 1994.

*Discovering Great Lakes Dunes.*
Elizabeth Brockwell-Tillman
and Earl Wolf. Michigan State
University Extension, Michigan
Sea Grant, and Gillette Natural
History Association. 1998.

*Dune Country: A Hiker's Guide
to the Indiana Dunes.* Glenda
Daniel. Swallow Press. 1984.

*Dune Type Inventory and Bar-
rier Dune Classification Study
of Michigan's Lake Michigan
Shore.* William R. Buckler.
Report of Investigation 23,
Geological Survey Division,
Department of Natural
Resources. 1979.

*Geology of Michigan.* John A.
Dorr, Jr. and Donald F. Eschman.
University of Michigan Press.
1970.

*Great Lakes Dune Ecosystems.*
Michigan State University
Extension. 1998.

*Grand Mere.* Grand Mere
Association. Kalamazoo Nature
Center. 1973.

*Michigan's Sand Dunes.*
Steven E. Wilson. Pamphlet 7,
Michigan Department of Natural
Resources, Geological Survey
Division.

*Michigan Coastal Dunes: A
Heritage Worth Saving.* Natural
Heritage Program. Michigan
Department of Natural
Resources. Poster.

*Sand Dunes of the Great Lakes.*
Edna and C. J. Elfont. Sleeping
Bear Press. 1997.

*Up and Down the Dunes.* Emma
Pitcher. Shirley Heinze Environmental Fund. 1987.

# SELECTED SCIENTIFIC ARTICLES

Arbogast, A. F., and W. L.
Loope. 1999. Maximum-
limiting ages of Lake Michigan
coastal dunes: their correlation
with Holocene lake level history.
*Journal of Great Lakes Research*
25: 372–382.

Cowles, H. C. 1899. The ecological relations of the vegetation on the sand dunes of Lake
Michigan. *Botanical Gazette*
27: 95–391.

Guire, K. E. and E. G. Voss.
1963. Distribution of distinctive
plants in the Great Lakes region.
*Michigan Botanist* 2: 99–114.

Hamze, S. I., and C. L. Jolls.
2000. Germination ecology of
a federally threatened endemic
thistle, *Cirsium pitcheri*, of the
Great Lakes. *American Midland
Naturalist* 143: 141–153.

Lichtner, J. 1998. Primary
succession and forest develop-

ment on coastal Lake Michigan
sand dunes. *Ecological Mono-
graphs* 68: 487–510.

Olson, J. 1958. Rates of
succession and soil changes on
southern Lake Michigan sand
dunes. *Botanical Gazette* 119:
125–170.

Thompson, T. A. 1992.
Beach-ridge development and
lake-level variation in southern
Lake Michigan. *Sedimentary
Geology* 80: 305–318.